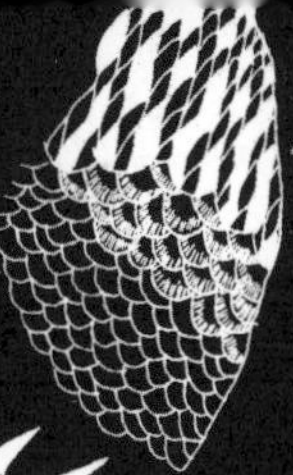

G000297618

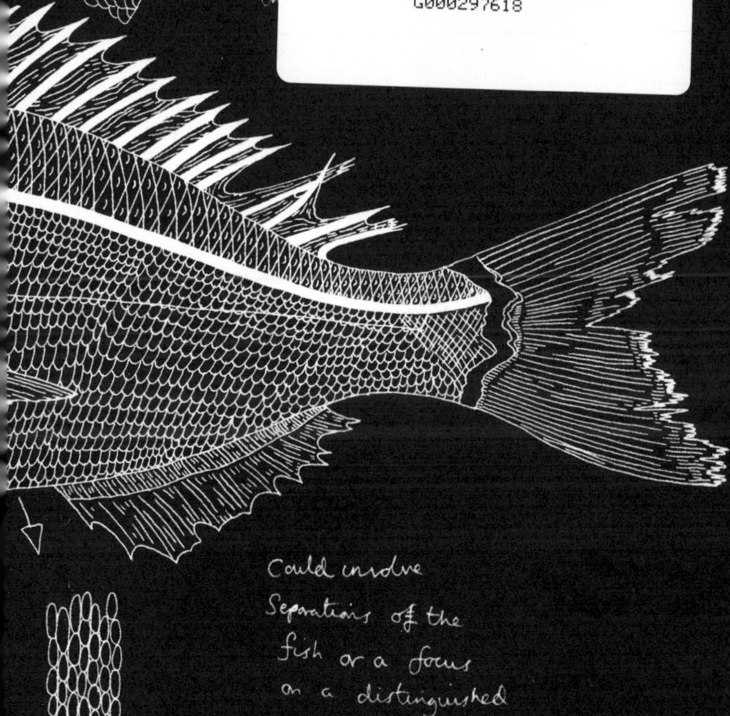

Could involve
Separations of the
fish or a focus
on a distinguished
part of the life form
e.g head, fins, etc.

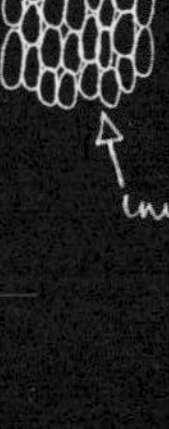

these
could have
a unique
unique scale   code of colour
formations   represented in
each fish/
life form

each fish could
be painted in
2 colours to make
a

full responses, lists of subjects
moments with each project including

would need to be a list of stones with interesting characteristics

SULPHUR

TYPES OF STONE INCLUDING STRUCTURAL COMPOSITION
- RUBY
- JASPER
- MOLDAVITE
- TIGER'S EYE
- AMETHYST
- SAND STONE
- ~~SULPHUR~~ SULPHUR
- HEMATITE
- CORAL

will also need to be a 'max' min size

REFERENCE POINTS
OWNED Book1 can also be a good reference point
- MY COLLECTION
- MUSEUM PHOTOGRAPH
- ONLINE PHOTOGRAPHS

METHODS
DIGITAL C
RED + BLU
RED + GR
- WATER COLO
- PEN (COP
- COLOURED
NEED TO BE
WHICH WILL
ACHIEVE.

CONTRASTS
acrylic inks would repro... interedar... placement on dry quickly
- HARDNESS
- STRUCTURE TYPE
- TOXICITY
- COLOUR

would need to have both can work of different Themes / ORDERS an opposite

colour you need can... using black

STAMPS
Could be used to make repeat pattern cover simple cover to reflects detailed illustrations inside the book.
book cover stone could be embossed and title written by hand

BASE COLOUR
WATER COLOUR
PENCIL COLOUR FOR DETAIL
OR
PENCIL BASE COLOUR
PENCIL / BLACK PEN DETAIL.

STOCK to use water color

"GEOGOGY STONES"

"FULL COLOUR OPTIONS"
water colour pencils could be used to add quality water colour & grunt to illustrations
Too achieve full colour stones would have to be kept ten actual size for book. AS MAX

could also be rendan in order

Could be used as a back cover vignette

FORMATS
STONE STRUCTURE
- Small format pocket sized book
- A5 - A6 size
- 48 pages
- 44 pages for illustrations
- rest for cov, index etc
- either single or double pages

→ see king penguin book Semi precious stones by N. WOOSTER 1952

HEMATITE TYPE -
GIVING illustrations FIG Numbers see references can be formed

THEME
BIRTH - E
PREDATOR - PREY
BONE - SHELL
MASK - SKUL
FEATHERED - SCALED
BIRD
FEATHER - BATWING
ARROWHEAD - SHARK TOOTH
PIKE REEL - FISH SKULL
FOSSIL - FISH
DINO - MAMMAL

OWL SKULL

INFORM THE VIEWER THROUGH DISCOVERY
paring need to be book, need to have a meaning / reference.

instead of figurative pattern... could be used.

Instead 1 can be separ

TRACING... than the stone size and shape

page number could change as well

CLAW

JAW
Objects would have more clarity with natural studies to avoid becoming repetitive.

SHARK TOOTH

NO DOMINANT THEME AT OF YET

text will reveal with real film microscope as well they could just reveal when microscope is used

BIRD EGG (SWALLOW TAILED KITE)

relatives items can be natural and manmade paired together if relevant.

illustrations don't c centrally or overlap single page would...

TEXT - BLUE... film... tests this.

bat wing

FEATHERED WING

ARROWHEAD

INUIT GOGGLES
CARAPACE
geology - a guide to semi - precious stones

...geology collection
...life of a museum

HIDDEN MUSEUM

A cabinet of curiosities

INTRODUCTION

*In the 16th Century, learned aristocrats curated collections of
wondrous, exotic artefacts across the fields of natural history,
anthropology and religion. These wunderkammer, or cabinets
of curiosities, were the precursors to museums
– tributes to the thirst for knowledge and understanding
that makes us human.*

*This book is a collection of specimens, geological samples
and historical objects. Behind each one hides something that
reveals a secret about its creation, evolution or connection to
history. Use the magnifying glass to explore the true nature of
each item in this little wunderkammer. I hope this book will
serve as a reminder of how rich and complex the planet we
live on can be. There are always worlds within worlds
– if you only care to look for them.*

SP

Amber is fossilised tree resin from prehistoric times. As it dripped down the trees it often trapped debris such as seeds, leaves and insects. The resin was then buried and exposed to high temperature and pressure over millions of years, making it solid and resistant to decay; a process called polymerisation. Most amber maintains the beautiful golden colour of the tree resin, but sometimes impurities cause amber to take on a bluish, cloudy appearance.

Mosquitoes evolved around 226 million years ago. The oldest known modern mosquito was found in a 79 million-year-old piece of Canadian amber.

Amber
&
Mosquito
(Culicoidea culicidae)

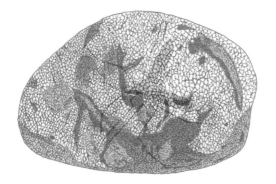

The priscacara was a bony perch fish that lived in the Eocene era, 50 million years ago. It had a flat, sunfish-like body and two large dorsal fins. The priscacara grew to a length of 35cm and swam in schools, feeding on a diet of crustaceans.

The yellow perch is its modern-day descendant. Native to North America, it has similarly large dorsal fins. Its body shape is longer and more oblong than its ancestor, but is also flat. It grows to around 25cm in length, swims in schools and feeds on invertebrates, crayfish and shrimp.

*Priscacara fossil
(Percidae priscacara)
&
Yellow perch
(Percidae perca)*

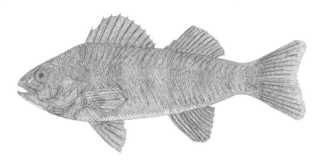

*Jet is a form of fossilised wood from the
monkey puzzle tree. It is a precursor to
coal, formed by a process of carbonisation
that occurred when trees in the Jurassic
period were washed into the sea and
buried under the seabed. Over millions of
years of pressure under the sedimentary
layers, the wood was transformed into
a resilient but still lightweight gemstone.
When polished, the black surface of jet
becomes shiny as a mirror.*

Monkey puzzle tree bark
(Araucaria araucana)
&
Jet

*Crocodiles can be differentiated from
their alligator cousins by their pointier
snouts, their protruding bottom teeth,
and the glands that allow them to
survive in saltwater environments.
Most crocodile species bury their eggs
in the sand, ensuring that the eggs are
kept upright. If they are flipped over,
the embryo can die by suffocating on its
own embryonic fluid.*

*The ancestors of crocodiles
were called archosaurs. They had
an unusually slow rate of molecular
evolution, and crocodiles have remained
largely unchanged in their physiognomy
for the past 200 million years.*

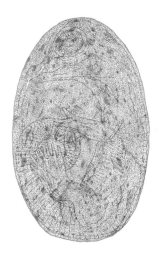

1

*Predator and prey are locked in an
eternal battle for survival. Each has
evolved highly attuned senses to ensure
the continuation of their species. Lions
have relatively close-set eyes, giving
them maximum binocular vision.
This means that both eye fields overlap,
making them good judges of distance
– an important part of stalking and
hunting. Gazelles, on the other hand,
like many grazing animals, have eyes
on either side of their heads, giving
them a wide field of vision, with only
a very small blind spot directly behind
them. This allows them to both sense
their predator and see where they
are running.*

Lion eye
(Panthera leo)
&
Gazelle eye
(Antilopini gazella)

Unlike a bird's wing, which is fairly rigid, a bat's wing has a flexible bone structure sheathed in a tough membrane called a 'patagium'. The skeletal system of bat wings is very similar to that of the human hand. Made up of four fingers and a thumb, which extends out of the wing, ending in a small claw, a bat's 'hand' is used not just for flight but for climbing trees and grasping food. When in flight, the flexible bones are moved in a circular motion like a butterfly stroke in swimming. The Latin word for bat is 'Chiroptera', which translates literally as 'hand wing'.

Bat wing
(Chiroptera)
&
Skeleton

Turtles are one of the oldest reptile groups on the planet, dating back to the Triassic era. Birds and reptiles share an evolutionary history evident in their scales (on the bodies of reptiles and the legs of birds), in their method of reproduction (the laying of eggs), and in their skeletal systems. Turtles and birds are the only two archosaurian animals today with a 'beak' made of bone and keratin.

The hawksbill sea turtle has a more defined beak than most turtles; its rounded shape is perfect for plucking at sea sponges. The hawk has a more radically hooked beak, made for tearing at flesh. Its skull is more streamlined and aerodynamic than the turtle's, with larger eyes set below the beak, awarding it excellent binocular vision.

Hawksbill sea turtle
(Eretmochelys imbricate)
&
Hawk skull
(Accipitridae)

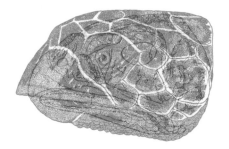

When moths and butterflies metamorphose from caterpillars into their adult beings, the process requires a complete deconstruction of their physical form. The moth weaves a cocoon of silk around its body and then sheds a hard layer of skin that becomes a pupa. Once safely inside this protective shell, the insect's hormones generate enzymes, which digest all the tissues except the 'imaginal discs', which hold the code to the caterpillar's end state. Once the tissues have been turned into protein, rapid cell division rebuilds the insect into its new physical configuration. This metamorphosis can take weeks or, in the case of swallowtail butterflies, years, as they only emerge when the temperature is right.

Cecropia moth cocoon
(Hyalophora cecropia)
&
Pupa

*Canopic jars were used by Ancient
Egyptians from 2000 to 1000BC to store
the internal organs of deceased nobles.
The stomach, intestine, liver and lungs
were each housed in individual jars
made of stone or pottery. The heart was
believed to be where the soul resided, so
was left inside the body. Each jar was
modeled after one of the funeral deities,
the four sons of Horus: Jackal-headed
Duamutef guarded the stomach,
human-headed Imsety was guardian
of the liver, baboon-headed Hapy
guarded the lungs and falcon-headed
Qebehsenuef guarded the intestines.
The removal of the organs prevented the
decomposition of the body during the
mummification process.*

Jackal-headed canopic jar
&
Stomach organ

A geode appears as a solid rock, but hiding within is a hollow core filled with crystals. A geode is usually an igneous rock, formed by cooling lava. Bubbles of carbon monoxide and water vapour get trapped within the lava, and when the lava cools, the gas dissolves, leaving a cavity. Sometimes a geode can be a sedimentary rock that has formed around an organic material such as an animal, wood or coral. When the organic material decays, this too leaves a hollow space. Both igneous and sedimentary rocks are porous. Over millions of years, mineral-rich ground water seeps into the cavities of the rocks, releasing chemicals inside the rock and depositing a further mineral crust. Very gradually these minerals build up to create crystalline structures, which can differ in colour and formation. A few common crystals found inside geodes are quartz, calcite, amethyst and hematite.

Geode
(Cryptocrystalline)
&
Quartz crystals

The magnolia genus has large, voluptuous flowers and deep, evergreen leaves. It is an ancient plant species with close ancestors in the Cretaceous. In fact, its modern iteration predates bees, and so it originally relied on beetles for pollination. Many of its attributes evolved in order to protect it from the beetle damage. It has 18 tough pistils and a distinctive, primitive seedpod. Cone-like in shape, with a woody texture, this seedpod splits open when it reaches maturity, revealing red, fleshy seeds, which hang from the centre on slim stems. These seeds are high in fat – good for providing energy to migrating birds, further spreading the dissemination of the plant.

Scarabs are the most diverse species of beetle with over 30,000 specimens. The Ancient Egyptians brought the sacred scarab to fame through their depictions in jewellery and art. This type of scarab is a dung beetle, which rolls dung into its burrows. The dung serves both as food and as the place in which the female lays her eggs. When the larvae hatch they consume the dung around them, emerging fully-fledged into the world. To the Ancient Egyptians, this represented regeneration and resurrection.

The scarab's flight wings are hidden beneath its thorax, protected under the front wings, called elytra.

Sacred scarab beetle
(Scarabaeidae)
&
Wings

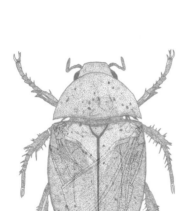

2

The Saxons ruled England from the
5th to the 11th Centuries. They used
a variety of weapons, many of them
exquisitely crafted. Helmets were the
rarest of all Saxon weaponry, used
only by nobles. The Sutton Hoo helmet
was discovered by archaeologists in
1939. It is beautifully made from
plates of iron fused together to cover
the cheek, neck and scalp. The iron
was then coated with shining copper,
which was embossed with scenes of
battle and heraldic animals. It dates
from around 600AD, and belonged to a
high-ranking cheiftain. It was found in
a grave along with other weaponry and
a 27 metre-long ship, which had been
dragged several hundred metres from a
nearby river.

Sutton Hoo saxon helmet
&
Human skull

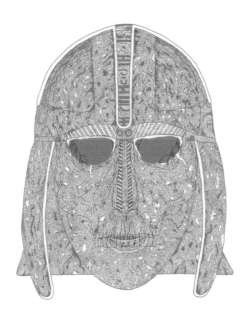

Known as buckeyes in America, conkers
are the seeds of the horse chestnut tree,
which was introduced from Turkey
in the 16th Century. They have a
rich, brown, polished surface with
a dull scar on the side, where they
were attached to the seed vessel. Their
seedpods are fleshy and green, covered
with spines. The seeds, especially the
young ones, are mildly toxic if eaten
raw. However, when boiled, they have
many uses including the treatment of
varicose veins and hemorrhoids and the
improvement of blood circulation.

The game "conkers" has been
popular in Britain since the mid 19th
Century. Two players thread their
conkers onto a piece of string and take
turns striking each other's conker until
one breaks.

Conker
(Aesculus hippocastanum)
&
Husk

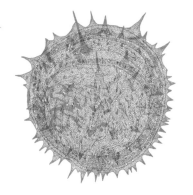

Latin for "remember you must die",
memento mori were symbolic reminders
of mortality; we are all the same in the
end, and death awaits us all. They often
took the form of jewellery or watches
(time being a symbol of impermanence)
and were popular from the 15th to the
early 20th Centuries. Early examples
featured literal representations of
death – skeletons, skulls and coffins.
Later examples were more abstract –
bats, snakes, willows and urns. After
the death of Prince Albert in 1861, the
trend for memento mori and memorial
jewellery grew to epic proportions.
This example is a Victorian memorial
locket, worn to commemorate a
deceased loved one. It features an
engraving of lilies (a flower that
symbolised death), and within it hides
a lock of the deceased's hair. Worn
close to the body, it was a constant
reminder of the fleeting nature of life.

Memento mori pendant
&
Lock of hair

*The Divje Babe flute is a broken shard
of what is thought to have been a
Neanderthal flute, dating from around
41,000BC. Carved from the femur of a
cave bear, it has two complete holes and
what may be the incomplete remains of
one hole on each end, meaning that the
bone may have had four or more holes
before being damaged. The artefact
is the subject of much debate. Some
archaeologists believe that it dates to the
Cro-Magnon era. Others believe that the
holes were pierced by a carnivore, and
not deliberately. However, it is unlikely
that a carnivore could have pierced such
regular holes. If indeed it is a musical
instrument, it would indicate that art
and creativity has been at the heart
of human existence since our genesis
as a species.*